Absolutely essential for anyone be
restoration project; a church is ab...
bricks...

Joseph Kelly
Editor, Church Building *magazine*

I warmly commend this visionary and excellent book to help people think about the use of their buildings. It is essential reading for a church thinking about extending its buildings and using its resources to be a faithful and effective gospel church.

Wallace Benn
Bishop of Lewes

This book charts the way through the kind of project that would test the resources and the resilience of the toughest congregation. From Chapter One, Julia Cameron shows how bricks and mortar can be used to serve the gospel. She doesn't gloss over the headaches or the heartbreaks, but remains insistently optimistic throughout. Some congregations would never contemplate a project like this; now they have an invaluable 'how to' tool to build for the gospel.

Richard Underwood
General Secretary, FIEC

Copyright © 2007 Julia Cameron

Published by **10Publishing**, a division of **10ofthose.com**.
13 Airport Road, Belfast BT3 9DY.

www.10ofthose.com

The idea for this new book grew out of a book previously published by
Paternoster Press in 1999 called *The church that went under*

The right of Julia Cameron to be identified as the author of this work has been
asserted by her in accordance with the Copyright, Designs and Patents Act 1988.

Unless otherwise indicated, Scripture quotations are take from the HOLY BIBLE,
NEW INTERNATIONAL VERSION. Copyright © 1973, 1978, 1984 International
Bible Society. Used by permission of Zondervan Bible Publishers.

All rights reserved. Except as may be permitted by the Copyright Act, no portion of
this publication may be reproduced in any form or by any means without prior
permission from the publisher.

ISBN 13: 978-1-906173-03-6

Design by: Jonathan Caplin **www.electrolyte.co.uk**
Print Management by: Print by Design Ltd **www.printbydesign.co.uk**

Royalties from the sale of this book have been assigned to IFES (see p71).
They will be used for student ministry in Cambodia in memory of Vivienne Curry,
who served as churchwarden in St Paul's, Howell Hill, Cheam.

BUILDING FOR THE GOSPEL
A handbook for the visionary and the terrified

Julia Cameron

Acknowledgements

It was Miles and Sara Thomson's idea that a book be written to help churches exploring the possibility of a major building project. Miles was the Rector of St Nicholas, Sevenoaks and its own story was published in 1999. Miles and his building committee wove their way through many difficulties; in addressing them they worked first of all to identify the biblical principles which related to the issues. This book seeks to look at those principles and to illustrate ways in which they can be applied.

Like hundreds in my generation, I owe a personal debt of gratitude under God to the Revd James Philip, for many years minister of Holyrood Abbey Church, Edinburgh. His systematic and pastoral preaching drew us into Scripture, and as we listened with our Bibles open in front of us, we found our minds and hearts dialoguing with the text. His wide-ranging illustrations, often from great works of literature we had not yet read, shed light on many passages. Here was eternal truth and it was compelling. It is a privilege to touch on Holyrood's story here.

I am grateful to Jonathan Carswell, Managing Director of 10ofthose.com, who saw the potential of this book when I first called him about it.

I have received help from a range of people; several are mentioned in the pages which follow but I would also like to thank Trevor Condy, Ian Ford, Nargiza Jedwab, Deborah Kelly, Philippa King, Richard Marshall, Jim Nelson and Jonathan West.

JEMC

Contents

Acknowledgements		4
Start here		7
1	Bricks, mortar, and the gospel of Christ	9
2	Churches: the same but different	12
3	Gaining permission is not easy	16
4	Children's work	19
5	Looking at the options	21
6	Believing God	27
7	Difficult questions	30
8	Setting and maintaining the tone	32
9	Building for the Gospel	34
10	A lot of money!	40
11	The exile	52
12	Moving back	56
'Now to him who is able'		60
Afterword by Angus MacLeay		62
Endnotes		64
Appendix 1 The art of belonging		66
Appendix 2 Lessons from Nehemiah		67
Appendix 3 Churches which would welcome visits		68

*With thankfulness to God for the lives of
Canon Miles Thomson, Tony and Eve Wilmot,
and others in and behind these stories who
have now joined the great cloud of witnesses.*

Start here

Churches in many towns and cities have struggled with decisions about their buildings over the past 20 years; even more will do so over the next 20 years. Church leaders have understandably seen reticence, even fear, when ideas of building projects have first been aired because they are so expensive.

For everyone finance is a major consideration. High property prices mean high mortgage repayments for those who can raise a deposit and high rents for those who can't. Christian parents find themselves pulled in several directions as their children want the same things their friends have; for single people the pressures are different, but they are no less real. It is part of our culture to live at the edge of our means, or slightly beyond that, so financial pressure has become part of life.

What can we 'afford' to give? How can we be sure that others will also give sacrificially, so as to make our own giving worthwhile? We all know that a church is more than a building, so do we really *need* better buildings anyway? Some questions are focused on the project itself, and others spring from a desire to protect our own or our families' financial interests from what could be perceived as a threat. It is good to ask questions.

Plans for a building project generally emerge out of reflection, discussion and strategic planning. In St Paul's & St George's Edinburgh, a further factor helped the church along that difficult stretch of road from 'Must we?' to 'We must.' I quote Emma Vardy, Project Director:

'We were helped in a prophetic way by Archbishop Kollini of Rwanda, who with no knowledge of our situation told the church that we must make the building bigger so more people could hear about Jesus.'

The Church in Africa also helped Reigate Baptist to come into being. Its pastor, John Bridger, was in Burundi in March 1995 at the invitation of David Ndaruhutse, leader of African Revival Ministries. As David prayed for John, then Associate Pastor of Redhill Baptist, he sensed the Lord's prompting to plant a new church. He shared this with the elders on his return, and as they talked and prayed, they believed it right to take it to the church membership. You will read the outcome in this book.

adventures of faith demand faith

Adventures of faith demand faith. And there can be times when that faith is not easy to maintain. The Christian life is often described as 'a fight of faith'. From the inside it can seem more like 'a fight for faith'.

Appendix 2, outlining lessons from Nehemiah, is taken from the writing of Miles Thomson, Rector of Sevenoaks, just a few months before he was diagnosed as suffering from a brain tumour. He went to be with Christ shortly after Christmas in 2000. 'Time and time again,' wrote Miles, 'turning to Nehemiah steadied our nerve and kept us to our vision.' He continued:

'We had, from the outset, committed ourselves not only to build a building, but to build a people; to build a church family who live under the authority of God's Word in God's world. It is still our prayer to keep on growing in faith as well as increasing in numbers, and to keep on training people in service, both in this country and overseas.

'We want to see more people finding a personal faith in Christ, and growing in their knowledge and love of him. We want to see more people taking Scripture as their guide at school or college, and in business, commerce, education or health; more going into pastoral ministry; more working to build God's church worldwide. We thank God for all the ways we are already seeing this happen; we pray he will enable us to carry on building for the gospel.'

May God give us that desire for 'more', so characteristic of the Apostle Paul, in all our churches.

I trust this little handbook will serve in three ways:

· to help churches already launching into building projects
· to encourage churches which are holding back
· to give ideas to churches not yet thinking about possibilities

To those ends I send it on its way with prayer.

JEMC

1
Bricks, mortar and the gospel of Christ

> All building projects have the same end in view: better facilities for teaching, training, fellowship, evangelism. This is the essence of church life: growth in depth, in numbers, in effectiveness, all for the glory of Christ.

We are getting much more audacious about building projects. Thirty years ago it seemed daring to spend a few thousand pounds on carpeting a church and replacing pews with chairs. Now we see a project in Edinburgh costing nearly £6m. To compare the number of zeros in those sums tells its own story.

Our buildings need to serve our purposes

While the expenditure has increased, the range of initiatives has also grown. Some churches have been built above shops; other churches meet in converted warehouses; a few have been designed in partnership with the local council or a local school. But the end is the same in each case: to provide a church building which is fitted to contemporary needs.

John Stott, the pastor theologian who was for many years Rector of All Souls, Langham Place in London's West End, summed up the needs of all human beings in three ways:

· everyone reaches for a transcendent presence outside the created universe
· everyone feels a need to matter in our digitalized and fragmented society
· everyone needs to know they belong when so little value is placed on family

If the first century Christian gospel is to relate to twenty-first century life, we must engage with people in these areas.[1] This book won't look at how we do that as a church, but at the kind of buildings we need to facilitate this. The two are closely connected. Our buildings need to be suited to our purposes on Sundays and for midweek events, but they are also the places where we equip Christians to be effective as the church at large: to shake salt and to shine light in our local community, through professional networks, and in public life.

Is it good stewardship?

We can naturally resent the need to spend money on bricks and mortar when the church is really about growing the Lord's people. Anglican church buildings don't belong to their members, so improving them can seem a bit like spending money on a house which you don't own. Isn't it better stewardship not to waste money like that?[2] Buildings belonging to Free Church congregations may legally be owned by their members, but even so, shouldn't we give that money to world mission instead? Again, these are good questions. This book is written for people with questions, so if you are asking them it is for you.

> our gospel is eternal; our buildings just a means to an end

Our gospel is eternal; our buildings just a means to an end. As more and more people are becoming Christians with no background in Scripture, it takes time (years, not weeks) for biblical thought patterns to be formed in them, and for a biblical worldview to be established. We need premises to facilitate that.

If you have read this far, there must be at least an inkling of dissatisfaction in your church. Dissatisfaction is a promising sign; it is like a seed which needs to be nurtured with prayer and godly vision. This is what will bring change.
You may want to build on a brown field site, if until now you have been renting school premises. Perhaps you just need to create more seating, and you're planning to replace a balcony which was, ironically, removed in the name of progress. You may be adding a church hall, or turning an unwelcoming hall into somewhere more comfortable, in keeping with our homes. Whatever your situation, you need money; and the church needs to have everyone on board to bring enough money in. Building projects will stretch a church. They will prove our faith, and they will strengthen faith.

In many churches, an evangelical ministry has over the years led to significant growth. As soon as a church starts to grow, and to see the potential for greater use of its buildings, that sense of healthy dissatisfaction begins to creep in. Members start to want something better for God's

> building projects will stretch a church. They will prove our faith, and they will strengthen faith.

service. This dissatisfaction may begin in the heart of the pastor, or of an elder, and then spread through the deacons or the church council. The journey from vision to completion will call for courage, perseverance and resilience; it will be demanding.

Hudson Taylor, the chemist's son from Barnsley in West Yorkshire who became a pioneer missionary to China, once said: 'There are three stages in any great work of God. Impossible. Difficult. Done!'[3] At the moment your building project may still loom as impossible.

2
Churches: the same but different

Whether you are in a market town or in a university town, in a farming community or in a multicultural city, your church acts as a beacon for the gospel in that context. Building projects vary enormously in cost, but acting as that beacon is the central theme of each and the unifying theme of all.

These brief outlines will give a feel for the churches featured in this book. Parallels between their situations and your own will not be hard to recognize.

St Paul's & St George's Episcopal Church, Edinburgh is in York Place, close to the city centre. As with all city churches, many members are professionals, but there are crowds of students, a growing number of retired people, and some who are unemployed. Its large building drew fewer than 20 members in 1985. Today P's and G's is a wonderful story of growth. Plans for a church plant in the city's expanding South East Wedge ran alongside the building project (Project 21). Total cost: close to £6m for a major refurbishment and additional Welcome Area, 2007.

Holyrood Abbey Church of Scotland, Edinburgh is about a mile east of P's and G's in Abbeyhill, at the junction of Marionville Road and Dalziel Place. Holyrood too draws an eclectic congregation from across the city. Its parish is made up mainly of tenements, housing a mixture of blue-collar workers, young professionals and families. Like each of the others, it is a church with a strong emphasis on world mission. Total cost: £1.6m for a major refurbishment and new hall in 2007.

St Andrew's Church, Leyland dates back to the twelfth century and serves a pleasant Lancashire town of 38,000 people where the main industry has been vehicle design since 1896. The first major reordering (the Vision Builders initiative) was completed in 1999 at a cost of £500,000. The second phase began in 2005. This included more refurbishment, and a new venture of faith: an evangelistic café-style church plant, meeting at 5 p.m. each Sunday in the local high school.

> **Reigate Baptist Church** dates from 1995. It is a charismatic church, and affiliated to the Baptist Union. By an arrangement with the local high school, the church provided its pupils with an all-weather pitch costing £250,000 in exchange for land on which to build. (An acre would have cost £1m on the open market.) The building itself, to double as a community centre, cost a further £1.5m - an ambitious project for a church of 100 members.
>
> **St Nicholas Church, Sevenoaks** is a medieval parish church surrounded by a graveyard, which limited the possibilities for an extension. The church is set at the north end of a largely comfortable commuter town. It took nearly 30 years for building plans eventually to come to fruition. In 2000, five years after the project was completed, a second morning service was needed; in 2006 the church planted a new congregation meeting in a local school. Total cost: £2m (1995) to create its Undercroft - halls beneath the church.

The story of St Nicholas forms the backbone to what follows, with illustrations added from other churches. God deals differently with different people in different places. There are often no single right answers to questions, as these stories show.

Early on in the process, each of the churches formulated its own vision statement to help focus the minds of the church family on the central reasons for the project.[4] Drawing up such a statement helps a whole church family to re-focus on what is essential. The wording varied in each case, but the message was the same. St Paul's & St George's communicated theirs with great effectiveness through FAQs (Frequently Asked Questions) on the Project 21 website. The touchstone question came first.

> **Q: How do the building plans fit with the vision?**
> **A:** Our mission is to bring people and God together; our vision is to build a Christ-centred, culturally-relevant, biblical community, worshipping and serving in the centre of Edinburgh; our strategy is to make, mature and mobilize people as disciples of Christ. [The answer then continued with specific references to plans.]

To make disciples, to nurture them towards maturity in Christ, and to equip them to serve him in strengthening his church - what a wonderful threefold purpose for buildings to serve. Are they fit for that purpose, now and looking into the future? This is the key issue.

John Bridger of Reigate Baptist circulated a thoughtful paper in 1999 entitled 'Church planting as an option for growth'.[5] To look at options and discuss them openly among the leadership and membership is a necessary foundation for the process.

> what a wonderful threefold purpose for buildings to serve

Those closely involved in all these projects have vivid memories of times when it seemed plain that God was on their side, and of other times when the discouragement of fellow church members was all they could hear; fellow believers who wanted the whole project forgotten for any one of a host of reasons. Because those church leaders pressed on, their stories - and the buildings which now stand - can exhort other churches for years to come.

I trust the accounts touched on here will bring inspiration, and strengthen the nerve of many other churches. Have a look at their websites or join them on a Sunday if you are in the area.

Getting the right people

Tony Wilmot gave a major thrust to the St Nicholas Undercroft.[6] He brought a mix of spiritual aspiration, shrewd judgment and professional experience. He was a man who looked at what was, and saw what could be.

The world over, able people are often the busiest, and church leaders can feel reticent about approaching those who are already heavily committed. On the basis that people have to make their own decisions before God, the policy Miles Thomson adopted was to invite those with suitable gifts and qualities to consider taking on responsibilities. He then trusted their judgment as to what they could manage. By approaching them, he was not putting any pressure on them. He knew they would reflect on the matter carefully and with prayer, and he felt they were the best placed to make the decision.

Church members themselves will be able to take the project to a certain level, but the church will then have to engage professional architects and an experienced project director. The project director may be found from within the church family, or move house to become part of the church family.

The wide-ranging role of project director will draw on every transferable skill a person can bring. This role is central to the smooth running of all aspects of the project: liaison with planning authorities and ecclesiastical authorities; interface with the work forces; and the flow of communication internally and externally. A critical part of the job is organisational ability. Emma Vardy brought this to Project 21 at P's and G's from a background in freelance work. As an event manager she had arranged professional conferences, concerts and other events; she also brought experience in fundraising. Brigadier Ian Dobbie, Project Director at St Nicholas, Sevenoaks, had served in the Royal Engineers and as an instructor at Sandhurst, and had held office in the Directorate of Manning in the Ministry of Defence, and as deputy chief of staff of an armoured division.

> the ethos of a project like this is very particular

The ethos of a project like this is very particular. It is not-for-profit like any charitable venture, but it is also for the glory of God. To be able to draw on the right people, who have skills and experience suffused by spiritual judgment, is all-important. For this reason, church members with professional backgrounds in finance, communications and construction are often brought onto a building committee so their expert advice can be pressed into service throughout the project.

How was the right combination of talent and judgment found in St Andrew's, Leyland? In God's providence, Steve Watson, an architect from Manchester, had moved into the town and joined the church two years before the Vision Builders project was launched. It soon became clear that he could serve as its consultant architect. He would sit in the church building to let the ambience sink in while he thought through ways of bringing the vision to its most effective and aesthetic reality. The church council accepted the tender from a local firm, Marland Builders and Contractors Ltd, and Jack Rimmer, its managing director and a longstanding church member, took personal responsibility for the project.

3
Gaining permission is not easy

As many readers will know from experience, householders can find themselves running into difficulties if they want to build an extension, or change the use of their home to incorporate a small business. We have words like 'red tape' and 'bureaucracy' in our language only because we need them! But for all the difficulties a homeowner may encounter, the matter is much more complex for a church.

As we have noted, Anglican church buildings do not belong to the congregation; church councils have no jurisdiction over them, though they are bound to keep them in good order. Permission to alter an Anglican building can be given only by the chancellor of the diocese on behalf of the Bishop. The chancellor himself is a layman, usually a former lawyer or judge.

> gaining permission can be a testing time

Outside courts have had this kind of jurisdiction over church buildings and their burial grounds for several hundred years. For independent churches, the decisions can usually be reached by the congregation, generally requiring a high percentage attendance at the meeting and a high percentage (if not unanimous) vote of those present for significant changes. All these decisions are then subject to town or city planning authorities in the same way that any other applications are.

Gaining permission can be a testing time, and those who are driving the planning applications need the prayerful support of the rest of the church. The processes can be long and complicated and in seeking an acceptable way forward church councils are rightly concerned to find solutions - and means of reaching them - which do not alienate local residents or businesses, and which commend the gospel rather than raise antagonism.

In Leyland the ffarington family, benefactors to the town of the lovely Worden Park, had a special interest in the church building. They had, generations earlier, paid for the Worden Chapel, set at the front of the nave. To keep this chapel in its original position would have given an asymmetric feel to the new interior of the church, and the church council applied for permission to move it further back. But the current generation of the family were naturally reluctant for changes to be made. The matter was taken to a Consistory Court, and the church lost its appeal. Eventually agreement was reached with the family and the chapel was moved back five feet. The ffarington family's gift is part of the long St Andrew's story, and no one wants to edit out these significant contributions in history. It was a happy outcome.

Permission for work to begin in St Paul's & St George's, Edinburgh was expected to take the statutory six to eight weeks, but this stretched to nine months. The church is a Grade A listed building and news of the plans was greeted with nervousness by Historic Scotland. This could have led to a great sense of frustration for the church family. However Dave Richards, the rector, boldly encouraged the church to press on with its giving programme, so that money given early could start to gain interest. These months were filled with much paper and millions of electrons as required information was gleaned and passed on; for most of this time there was little hard news to report Sunday by Sunday. Dave, Emma Vardy and the building committee had to accept the delays and pray for grace in all their dealings with those who were causing them. While it was generally assumed that permission would be granted, all who gave money at this stage were exercising faith. Eventually, with the support of the Scottish Executive, Edinburgh City Council approved the plans, and Historic Scotland's concerns were overruled. By the time the church moved out of its building, three-quarters of the necessary funding had been pledged.

> this could have led to a great sense of frustration

For St Nicholas, Sevenoaks (or St Nick's as it is often known) the waiting period was much longer. It took 30 years to gain all necessary permissions for its Undercroft. The Sevenoaks Society, the Royal Fine Arts Commission and the (then) Department of the Environment had each mobilized support against the church. Over this protracted period two different solutions to the accommodation problem were ventured but rejected. The level of antagonism grew so high at one stage that Michael Heseltine, then Secretary for the Environment, involved himself personally and ruled against the plans.

The words 'but God' sum up much of what happened in Sevenoaks over the next few years, as he worked to change situations, and to change people's minds, in answer to prayer. The solution to gain eventual acceptance was to be much more radical. It might appear that the secular authorities overruled the plans in the interests of the publics they served. Perhaps the higher truth was simply: 'But God meant it for good'.

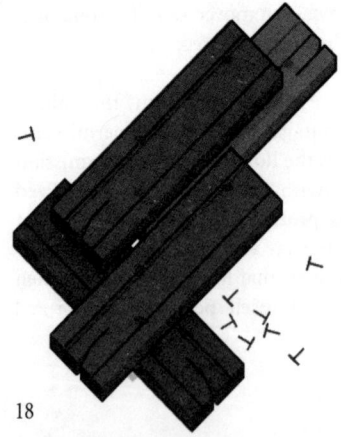

4
Children's work

Spiritual and practical considerations tend to run in parallel; indeed the spiritual solution can prove to be the practical one. One area in which this is often seen is how arrangements work for children's clubs on Sundays.

A large proportion of Christians are converted when they are young, through the influence of Sunday clubs, through school friends, camps or house parties, through reading the Bible for themselves, or perhaps through a university mission. As people start work and take on mortgages, spiritual matters are often sidelined. There is no social pressure to have a baby 'christened', so the first big opportunity for parents to be linked with a church can be when their children reach Sunday school age. Many parents still want their children to learn about God. They may feel inadequate themselves as teachers if they do not have a good grasp of the Christian faith. What they need is a church that will welcome the family, and make them feel at home.

> premises which serve well will always be a basic need

The location of the crèche and the Sunday clubs is critical. Is the accommodation warm, safe, light? Will parents be able to leave the children and get into the service in good time or will they have to walk some distance from the crèche or Sunday clubs to the main church building and risk arriving late, and feeling conspicuous?

Paul Batchelor, treasurer of the *Building for the Gospel* project in Sevenoaks (or B for the G as it came to be called), was drawn back to church through its children's work. He and his wife, Janet, had both attended the University Church while students in Cambridge, but it was not until they started bringing their children to the Sunday school in Sevenoaks that they found personal faith in Christ. Paul Batchelor was at that time on the international management team of Coopers & Lybrand (now PricewaterhouseCoopers). Very soon after he came to faith, the then building fund treasurer resigned, and Paul was asked to replace him. 'I had a crowded diary, heavy travel commitments, and knew it could only bring more pressure. But after much discussion and prayer, Janet and I decided I had the needed skills and it was right for me to accept. For both of us it proved a faith-building exercise, and we were privileged to be part of it.'

Up and down the country we are seeing more and more churches with good premises taking creative initiatives in family evangelism during the week. Midweek clubs for pre-school children can be a highlight for mothers at home with their little ones, and for nannies. Here the children can play while parents and carers talk. Often there is some singing and a Bible story. The children enjoy learning the songs and spiritual truths can start to seep in through the words. With such little biblical knowledge taught in schools, adults may be as new to the Bible stories as the children.

With lower emphasis placed on family life, and with Sunday as an alternative shopping day, these midweek events have enormous potential for the light and salt of the gospel. They can spark off events for fathers, too, through initiatives like 'Just Daddy and me' (where fathers are invited by their children to come on an occasional Saturday to see what the children are doing), or men's breakfasts.

A few creative members of a church family can design an imaginative programme for a year, building on the Christian calendar and Christian values in the home and the public square. But premises which serve well will always be a basic need.

5
Looking at options

Jesus used a building project to help illustrate what it means to become a disciple. Just like people who are about to embark on a major construction, he said, anyone thinking hard about being a Christian also needs to count the cost.

Suppose one of you wants to build a tower. Will he not first sit down and estimate the cost to see if he has enough money to complete it? For if he lays the foundation and is not able to finish it, everyone who sees it will ridicule him, saying, 'The fellow began to build and was not able to finish.' (see Luke 14:25-35)

Jesus knew people's hearts. He knew that matters which touch our bank accounts are taken very seriously. And that is why he chose the illustration.

For the church to look honestly at options, each has to be weighed carefully. Whatever the way ahead, the cost must be counted with care. A feasibility study will be needed, presenting the hard data of what things will mean in cash terms.

Three main options

There are likely to be three main options. These of course will vary. A church with a small footprint, perhaps defined by the convergence of two roads like Holyrood Abbey Church in Edinburgh, may have less choice as to how plans can proceed. But no church has fewer than two options, and most churches have three.

Option 1 This is always the same. It is to do nothing. To proceed in this way is a real option, and needs to be set alongside the others as a possible choice. What are the benefits of doing nothing? What would be the longer-term costs?

It is critical for this option to be examined carefully as the discussion will determine much of the way a church proceeds.

While a cost/benefit analysis of any option can be handled on a flip chart at a church meeting, the benefits and deficits of the 'do nothing' option can also be discussed at home and with friends, for everyone knows the current buildings well. If there are problems about accommodation, this option will pass the problems on to those who follow. While it demands no courage or sacrificial giving, it offers no solution to pressing needs.

> **Option 2** This is the relatively low-cost way forward. It will vary from church to church.
>
> **Option 3** This is the visionary way forward. It will mean a level of sacrificial giving which will hurt everyone: the low earners, the middle earners and the high earners.

In the case of St Nicholas, Sevenoaks, Option 3 was to dig a hall beneath the church. It meant spending a lot of money simply to explore its technical feasibility. In the 1990s no hall had ever been attempted beneath a medieval building anywhere in the world. It was, as the architect Robert Potter later described it, 'almost too challenging to do more than whisper'.

Let's think behind the scenes. There was an unavoidable question in the back of everyone's mind. 'What if exploratory digs show the project is not feasible?' Money was needed for the digs, and it could all come to nothing as the fourteenth century foundations might not allow a substructure. To brace a church for generous giving with a goal in sight is one thing; to ask people for significant giving when the goal is uncertain is another.

West Street Baptist in Dunstable was in a similar situation in 2000; it needed money for a project which might not work out. The plan was to purchase The Plume of Feathers, an ancient coaching inn, well known in the town, and just three doors away. This was a derelict Grade II listed building which dated back to the 1700s, and permission for change of use was by no means certain. Would members of the church family take a godly risk and make a blind investment? They did, and they kept giving as new problems kept being discovered. The inn had fallen into such disrepair that seven tonnes of rubbish needed to be removed from it. Refurbishment began late in 2003 and The Way, a Christian Community Centre for the town, was opened in October 2004. Thank God that members of West Street took that risk. Now, partly as a result of its ministry through The Way, the church is looking at a second project: extending its own premises.

> some initial blind investment is almost always necessary

Some initial blind investment is almost always necessary. Similar circumstances (i.e. of requiring a substantial sum with no assurance of the project's success) could arise in any church. Perhaps a nearby building comes onto the market, as happened in Dunstable, and it needs to be purchased before permission for change of use is certain. Or adjacent land needs to be bought quickly before planning permission for an extension can be worked through.

Or even if the church has land on which to build, suppose complex (and therefore expensive) plans need to be drawn up to submit to the town council or ecclesiastical bodies. Which church members will provide this money? It takes risky faith to sacrifice a family holiday or a change of car at such an early stage of a building project. No one doubts that.

Moving up the option scale

The original plans at St Andrew's, Leyland had been costed at £370,000. Steve Watson's advice to replace the church ceiling at the same time took the total to nearly £500,000. The Future Dreams adopted in 2005 included plans to take out the pews and replace them with chairs, and to install small TV screens below the galleries and a large plasma screen at the front of the nave. This plan would enable those without sight lines to see the preacher, and ensure that everyone could read words on a screen. But the screens alone would cost £16,000. Another major dream lay beyond that: to rebuild the large adjoining church hall, and to create a new role for the staff team - a leader for a 5 p.m. café-style service. So substantially more money would be needed as time went on. Option 3 was being stretched; faith that God would provide needed to be stretched too.

With 80 members at a parish weekend away, the church still seemed full. Regular members had to be asked to meet in the church hall for the annual Remembrance service. A local soldier had died in Iraq and the church would be jam-packed on Remembrance Sunday for years to come. The Vision Builders were looking to God for further growth. How could they not grasp the opportunities he was giving?

The church family in Holyrood Abbey, Edinburgh was to make the same move. Here plans had been drawn up for a refurbished sanctuary and a brand new hall with space for a church office. Holyrood has around 260 members and during their pledge month they had raised £843,000. Together with proceeds from the sale of a property, this would comfortably achieve the envisaged £1.2m needed for their project. But very soon a major obstacle appeared. Philip Hair, the minister, tells the story:

Heartened and encouraged, we sent the plans out to tender. One can only begin to imagine the shock and consternation when the lowest tender came in at nearly £1.9m. Prices, especially that of steel, had rocketed in the intervening months. We also discovered unforeseen technical difficulties.

Several weeks later we presented four options to the congregational board (the committee responsible for the fabric and financial matters). After much discussion the board voted with a large majority to opt for a scaled-down version of the new hall. There would be no provision for an upper hall or for office space, but the original budget of £1.2m would remain unchanged.

Soon afterwards Phil Hair began having sleepless nights. He became more and more convinced that the congregational board had not come to the right decision. The new premises would be good, but not give the flexibility needed for evangelistic initiatives - the whole reason for the project. In short, it would be 'Option 2' rather than 'Option 3'.

Phil consulted with several of his senior elders, expressing his growing concern that the board had taken the wrong decision. The elders advised him to call a special board meeting and spell out why he felt it was so important that members revisit the matter. Holyrood does not have a rich congregation. To achieve the extra accommodation which Phil sensed was needed would cost a further £400,000. Many had already given sacrificially, almost beyond their means.

Over to Phil again:

The day of the meeting arrived. Humanly speaking, World War III should have broken out, but it didn't. The discussion was marked with understanding and everyone spoke wisely and graciously. After two hours there was still no consensus, with members almost equally divided as to what to do. A break was needed. After the break we sought the Lord's will again in a time of prayer. Within about fifteen minutes we moved to a vote. It was unanimous. We would proceed in faith with the £1.6m option.

That vote brought with it a sense of peace that the additional funds would be raised. On the following Sunday Phil explained matters to the congregation, laying out the need and inviting people to give, but without exerting pressure. The board had moved forward with a sense of godly anticipation and the Lord would honour this. The members' response was not long in coming. Two weeks later the church family rose to sing 'Praise God from whom all blessings flow' and 'Great is thy faithfulness'. A further £250,000 had been pledged.

'We did it - you can too'

St Nick's was greatly helped by what All Souls, Langham Place had done years before. Being urged on by a church which has already been through similar difficulties can be a huge help. The churches on p68-69 are just a few of those who have already embarked on building projects. Linking with a church for prayer, partnership and encouragement can be a wonderful support. St Nicholas looked to All Souls in this way. Bishop Michael Baughen, its former rector, came down to St Nick's at a critical period, when it was vital that the whole church family gain ownership of the plans.

He related a story which gave heart: All site meetings at All Souls started with prayer. The workmen had felt uneasy about this in the first few weeks, and had perhaps been a little embarrassed. But they got used to it, and from time to time even asked the church to pray when they found something problematic. One afternoon the site manager asked for prayer as the local district inspector was being difficult. That evening at the church prayer meeting, the matter was again brought to the Lord; people didn't know what to ask, but were keenly aware that this man was causing problems. The next morning, news came that he had been promoted. Bishop Michael concluded, 'He was delighted and so were we.'

Michael Baughen again raised the question of money. Is it right for a church to spend so much on its buildings? He spoke perceptively. Some who ask have genuine concerns, and need reassurance about the project. If people prefer to make a substantial gift to overseas work instead, they must feel free to do that. But it would not be right to think of a building project as 'spending money on ourselves'. It is 'spending money for Christ's kingdom'. And as homes become more and more comfortable, it is appropriate to think of ambience for a church, too, so friends invited to church feel at home. Churches, he added, will always testify that giving to a building project releases money for other things, and general giving moves up significantly.

> giving to a building project releases money for other things

Bishop Michael also helped St Paul's & St George's. He had followed its story since 1985 when, large and empty, it was taken on as a church plant by a congregation in Corstorphine, four miles west of the city centre. Dave Richards knew of this link and invited him to preach one Sunday, and to meet with church leaders. As he had done in Holy Trinity Platt, Manchester in the 1960s, in All Souls, Langham Place in the 1970s, and more recently in Sevenoaks, Bishop Michael urged the church family to look forward in confident hope. Taking the congregation into Exodus 33 he exhorted them to plead just as Moses had: 'Lord, teach me your ways, and show me your glory.'

He urged people to think of generations to come. 'It gives me joy,' he said, 'to see All Souls crowded, used and working, and to think of what it's done for hundreds and hundreds and *hundreds* of young people.'

6
Believing God

An adventure of faith

Hudson Taylor believed that 'God's work done in God's way will never lack supply'; and these words have strengthened many Christians in faith ventures ever since. Surely that 'supply' does not relate only to finance. It also applies to the spiritual, physical and emotional resources people need to keep going if they are to finish the task. As Hudson Taylor prayed for new workers, he never imagined the way ahead would be trouble-free. His team of recruits in China, without any formal language training, was derided in the House of Commons as 'the pigtail mission'; and there was homesickness and lack of trust among those who found it difficult to accept his authority. Years later, just as he was to hand over office as general director of the mission, his chosen successor died - with over 50 other missionaries - in the Boxer uprising. Hudson Taylor's enduring faith was not born out of naivety.

For any church venturing out in faith, God is at work in people's lives - at times remarkably; at other times almost imperceptibly. Stickability is needed for a building project. From the first pledge day to moving back into the church could typically take as long as four years and throughout this time the initial effort needs to be sustained.

How will local historians chronicle hopes and plans for your church in decades to come? That will depend on who writes the history. To onlookers, these projects consist only in bricks and mortar, permissions and contractors. But that leaves too much unsaid. The whole purpose is to create facilities to bring honour to Jesus Christ. They are spiritual projects, and can be truly perceived only with that purpose in view.

The Apostle Paul wrote to the Ephesians that God is 'able to do immeasurably more than all we can ask or imagine' (Eph. 3:20,21). Bishop Michael Baughen had referred to this on his visit to St Nick's as a 'very major promise', and these rich verses were to prove a source of hope and stability throughout the whole process. The same verses were included at the end of a longer passage (vv. 13-22) read by Steve Watson at the celebration service in St Andrew's, Leyland.

If Scripture is God's Word for his world for all time, and if it is true and sufficient, then we will be able to draw strength from it. Several churches have found renewed strength for a building project through turning to Paul's letter to the Romans. In the opening verses of Romans 5, Paul links suffering with perseverance, character building and hope. We are justified; we have peace with God. With this foundation, difficult or painful experiences can be put into a broader perspective - the eternal dimension.

For St Paul's & St George's Church in Edinburgh, different verses had special significance. One was Joshua 3:4: 'Then you will know which way to go since you have never been this way before.' Following God often means stepping out into the unknown. Dave Richards preached a series of sermons on trusting God, and invited John Ortberg (author of *If You Want to Walk on Water, You've Got to Get Out of the Boat*) to preach as part of that series on Jesus' invitation to trust him.

Even when the plans remain the same, rising costs are not unknown. As the cost rose for St Paul's & St George's, the church family turned to the life of Abraham, who was promised a son when humanly speaking this was impossible. 'Yet he did not waver through unbelief regarding the promise of God, but was strengthened in his faith and gave glory to God, being fully persuaded that God had power to do what he had promised.' (Rom. 4:20,21) As Emma Vardy, Project Director, explained, 'It challenged our ability and willingness to trust God despite current events.'

Reigate Baptist Church started life in a community centre on a large housing estate. The centre was, to quote John Bridger the pastor, 'completely unchurchy' (which was in its favour for drawing in new families) but it soon became too small. The church then moved to a local school hall. Through the five-year journey from the church's founding in

1995 to the opening of its current building, John Bridger turned over and over again to God's promise to the Israelites, as revealed in a dream to Nathan: 'I will provide a place for my people Israel and will plant them so that they can have a home of their own and no longer be disturbed.' (2 Sam. 7:10)

The Lord brings sustenance through deepening our grasp of his character - the character of the One who is able to do more than all we ask or imagine - and through reminding us of his dealings in the past, which are recorded for our learning.

Keeping everyone in touch

Any large project involves simultaneous action on several fronts, and careful coordination of each of the different areas of activity. While the elders or church council retain ultimate responsibility, much of the ongoing work is usually delegated to a building committee, a finance advisory group with deep experience in banking, finance and legal matters, and a Project Director. These people all shoulder a heavy load. In addition some churches hire the services, whether as a consultant or as short-term staff, of a professional fundraiser.

It can be a great encouragement to these people to know that the church family is keeping in touch - reading minutes of planning meetings etc.

Holyrood Abbey Church along with St Paul's & St George's used their websites to build an online archive of photographs of the work in progress. The P's and G's site included a virtual walk-through of the new Welcome Area, then yet to be built. This area forms a significant feature of the renovated church, situated on a major bus route in the city centre.

On the wider communications front, local papers and news sites can keep the public informed of progress. Building projects costing millions of pounds raised from within the church family will naturally become a topic of conversation in a town or suburb.

7
Difficult questions

Churches will sometimes have to address difficult, even painful questions which divide the congregation. This is not always the case, and even for a church the size of St Paul's & St George's, it did not happen. For St Nicholas a matter which raised particular problems was the future of the Campaigners work.[7] The fact that the church council had on several occasions declared its commitment to preserve this did not seem to allay fears. The difficulty arose from a need to sell the hall where the Campaigners met.

> churches will sometimes have to address difficult, even painful questions

Surely God would lead in this, as he had in everything else. Yet there seemed no obvious answer. The division of views was unhelpful and was hindering Christian fellowship among some. Comments and questions came in writing, in telephone calls, and in conversations with the church staff and with church council members.

In a last effort to win the confidence of those who were disaffected, a final open meeting was planned for questions and answers. After this, the matter would be considered closed. It turned out to be a good meeting, with just over 20 people present, and everyone having the chance to speak and to feel they had been heard. While members of the building committee fielded the questions, their wives met to pray. At the end of the evening an envelope was handed in - a new financial pledge to replace one which had been rescinded earlier. The Holy Spirit was at work changing hearts, changing minds, changing attitudes.

No one doubted the vital and distinctive role played by Campaigners in the life of the church. A considerable amount of effort had already gone into scouring the town for a suitable alternative venue. A boys' school had been a possibility. It would have given enough space in halls and classrooms, but the Campaigner leaders felt, understandably, that being spread around the building would lose a sense of cohesion. Use of the school would also have incurred considerable expense. But this would have been offset by the loss of maintenance costs for the hall, which were eating into church funds.

A final decision on the sale of the hall was to be taken by the church council the following week. The outcome was awaited with more than a little interest.

On that morning, Miles and Sara Thomson were reading Proverbs 21, which opens with a strong affirmation of God's sovereignty: 'The King's heart is in the hand of the Lord; he directs it like a watercourse wherever he pleases.' Christians can sometimes let that certainty of his ultimate control become feint in their thinking. Here was another reminder that everything was in his hands. It was unanimously agreed that the hall be sold. Such unanimity was a precious thing. The way ahead for Campaigners was still not clear, but the confident vote of the church council reflected their reliance on a sovereign God.

Many churches in the country which have moved forward in a project like this probably have a similar story to tell of understandable concern in relation to one group or another. To hear how each of those situations was resolved would be faith-building. In St Nicholas the dénouement was entirely unexpected, and surely an answer to people's prayers. The minister of the United Reformed Church made an approach, asking if the Campaigners would like to meet in his building. This was close to the leaders' own home, and provided excellent hall space and a much better kitchen for use by those working for cookery badges. Yet again a solution had been found which was better than people could have hoped for or imagined.

8
Setting and maintaining the tone

On the principle that the leaders should lead, members of the church council (or the elders and deacons) are often invited to make their pledges two weeks before the main gift day for the congregation, and the total pledged is announced on the Sunday in between. This can set a tone and demonstrate that the leadership is united in its commitment.

On the first gift day in St Nicholas, people placed their envelopes in the boxes as they arrived for the service. After they had been brought to the front of the church, and received with prayer, Miles Thomson preached. The passage was from Philippians 3, which contrasted any merit on this earth with 'the surpassing greatness of knowing Christ Jesus'. Miles simply expounded what Paul wrote.

The evening service followed a similar pattern, with Johnny Juckes (now Rector of St Andrew's, Kirk Ella in Humberside) preaching on Isaiah 40. The chapter brought a powerful reminder that God can be trusted. Halfway through the service Colin and Hazel Maunsell, on a brief home assignment, were interviewed about their work in Ethiopia. It was fitting to express commitment to them on this particular Sunday. The building project could not be allowed to preoccupy a missions-minded church or to divert resources from missions.

Sunday by Sunday, the teaching of Scripture addressed issues in people's minds. That summer Miles was preaching a series on the Songs of Ascent - the psalms sung by the people of Israel as they travelled to Jerusalem for special feasts. They were psalms of pilgrimage from which strong parallels could be drawn. In Israel then, as in Sevenoaks, a group of God's people were on a journey. By the end of July Miles had reached Psalm 133 with its clear plea for unity. Unity in the gospel was of far greater importance than anything else. Whatever people felt about the Undercroft, St Nick's must guard its unity.

Tony Wilmot was not only the driving force behind B for the G, but also Chairman of the mission committee; he had spent much of his life in

Africa and could not ignore the pressing needs of the developing world. In August that year he preached with authority on Jesus' own words: 'The poor you will always have with you' (Mt. 26:11). Here, he said, was an unusual opportunity to give sacrificially for God's glory. Again it was the teaching of Scripture being applied to the questions of the moment.

If the Bible is our guide for all that we believe and all that we do, it has to remain in the forefront of our thinking. That is what it means to be an evangelical church. And the preaching on Sundays and daily Bible reading in homes play a very significant role in keeping projects on course, in proportion, and - all importantly - focused on their spiritual goal.

While church leaders work diligently at maintaining unity, it would be naive to think this is always possible in our fallen world. So while some in the church family go without a holiday, make other sacrificial choices, or take on extra work to be able to give more to the project, a few may leave the church, feeling a big mistake is being made.

Those who feel the money should be going overseas could be invited to give to the mission fund rather than to the building project. But if in the end people resolve to leave and to worship elsewhere, then that decision, though sad, has to be respected. They should be assured of a welcome back if at any stage they feel it right to return.

> if people resolve to leave and worship elsewhere, then that decision has to be respected.

Life would certainly be far easier without any building project at all. Again and again it is Scripture's central thrust that keeps churches going: the thread traced from Genesis to Revelation of a missionary God with an ever-contemporary message for everyone to hear.

9
Building for the Gospel

Time for an interlude before the money chapter!

B for the G is, in a sense, what all church life is about. The Apostle Paul wrote to the Corinthian Christians about building for the gospel in their own lives:

By the grace God has given me, I laid a foundation as an expert builder, and someone else is building on it. But each one should be careful how he builds. For no one can lay any foundation other than the one already laid, which is Jesus Christ. If any man builds on this foundation using gold, silver, costly stones, wood, hay or straw, his work will be shown for what it is, because the Day will bring it to light. It will be revealed with fire, and the fire will test the quality of each man's work. If what he has built survives, he will receive his reward. If it is burned up, he will suffer loss; he himself will be saved, but only as one escaping through the flames. (1 Cor. 3:10-15)

Leaders of all groups and activities are, all the time, 'building' in the lives of their members. God has given special gifts to some in the church so that, together and individually, the whole church family can be built up. In his letter to the Christians in Ephesus, Paul was clear about his aim: '…until we all reach unity in the faith and in the knowledge of the Son of God and become mature, attaining to the whole measure of the fulness of Christ' (Eph. 4:12,13). This must be at the heart of any community of Christians.

B for the G beyond our church

Church life can sometimes be perceived by its members simply in terms of 'us and God'. The call to be salt and light in society, and to bring the good news of the gospel to friends, neighbours and colleagues, can slip off the edge of Christian thinking. The whole earth belongs to God, and Christians are in a sense claiming it for him. This applies in different ways to teachers, businessmen, medics and so forth, and everyone needs help in learning how to apply biblical principles in their work. A local church may not have specialists in all these areas, but leaders can encourage the whole church family to 'think Christianly', and point people in the right direction for stimulus and guidance.

Through praying, we can also build for the gospel in the lives of people we have never met. Missionaries from our churches are engaged in a wide range of activities across the continents: literature ministry,

> through praying, we can also build for the gospel in the lives of people we have never met

church planting, mission personnel work, student evangelism… Some serve in the UK offices of overseas missions. Whether on the front line or in the back room they are all working to strengthen Christ's church around the world. And as church members get behind them in prayer, they are themselves building for the gospel in villages and small towns, on university campuses and in the world's mega-cities.

Thinking Christianly

A casual observer would be forgiven for thinking Christianity out of place in the postmodern world: superseded, redundant. A more reflective person might wonder how the whole basis of modern Western civilization could have been so quickly overturned.

In the mid 1980s John Stott coined the phrase 'double listening', saying Christians must listen to the Word (the Bible) and listen to the world. Then we will be able to relate one to the other, and engage with the questions which people are asking. Preaching and teaching in our churches must help Christians to do that. The idea of an objective body of truth has been completely eroded in our highly individualistic and relativist culture. It might seem quaint to think of the Bible as our authority for what to believe and how to live; Jurassic Park may seem the best place for preachers.

Relating the Word to the world is a learned skill. Building for the gospel must include the task of building in people's lives to equip them as Christians in the here and now: as Christian parents, neighbours and friends; and as Christians in the workplace. Let us never underestimate the wholly different mindset of the Christian. It takes time for people to absorb Christian values; a Christian worldview cannot be microwaved. We must work hard at helping one another to learn, love and obey Scripture. We have moved a long way since the 1960s in attitudes to authority, and Christians can easily lose a sense of Scripture being our authority - our authority on what is true, and our authority on how to behave.

As we have already noted, this book is not about how the church must do that, but only about the premises needed for it to be done.

> we can easily lose a sense of Scripture being our authority

Commuting is a feature of life in satellite towns and suburbs around all our major cities. The sheer numbers who pile off the trains every evening, and often don't get home until late, mean midweek church life has its limitations. Churches need to get behind people in the hard places of business and commerce and pray for them.[8] Tired commuters have little enough time with their families and may find home groups or other evening meetings just too hard to get to. Different ways need to be found of helping them to 'dialogue' with the Bible for themselves, to interrogate it, to prove it as a trustworthy and sufficient guide, both for the big questions in their professions and for everyday living.

Even before the St Nicholas Undercroft was completed, it was agreed in principle that the next step would be to recruit an additional member to the church staff. This person would get alongside the many leaders (in the children's work, youth work, home groups, weekly meetings etc) to equip and resource them, provide whatever on-the-job training and support they needed, and keep a constant eye open for those with overload. He would also nurture a future generation of leaders. The job title was eventually agreed as Director of Pastoring and Training. It was essentially an enabling job, a bit like that of a team coach. Someone nicknamed it 'church bodybuilder'.

Philip de Grey-Warter (now Rector of Fowey) joined the staff in this role. Buildings are needed for effective ministry, but having the right people in the right jobs to enable that ministry is just as vital. With Philip on board, two major new initiatives became possible.

Leadership training

First he and the other clergy were able to look out for emerging leaders in whom time and training should be invested. Phil and his wife, Naomi, ran a training course for groups of six or eight at a time, culminating in a residential weekend. Over the course of a programme like this, the members get to know one another, and there is a team spirit of learning together. 'The aim,' said Philip, 'was to enthuse and envision members for biblical ministry and spiritual leadership, and begin to equip them for it.'

There was no pressure on anyone to take up a leadership role straight away, nor any pressure on Phil to put someone in a gap which needed to be plugged. It is vital to have people in the right roles, whether working with children, teenagers, home groups or in any other capacity. As leaders, these people are themselves building for the gospel in the lives of others.

> these people are themselves building for the gospel in the lives of others

Toolbox

Secondly there was Toolbox. This is now used in different forms around the country. On an occasional basis, evening services are replaced by seminars on a range of subjects. These seminars look at the practical aspects of what it means to be a Christian in the world of business, commerce, and education; in the family; in the community. It would be misleading to say there is no evening worship on those Sundays, for the whole evening is an act of worship, as church members seek to work out how the Lordship of Christ affects their lives.

Good, flexible, modern premises give openings for many different ways of engaging the minds and hearts of Christians and of those who have not professed faith. Toolbox, lunchtime meetings for those working locally, pre-school groups, 'Just Daddy and Me', retired men's lunches, women's evangelism, apologetics seminars - our aim is surely to nurture the life of Christ in people, the life of the One through whom everything was created, and in whom all things hold together (Col. 1:17). In a world of fragmented and disintegrated thinking - disconnected from the present and disconnected from the past - to be reminded of things 'holding together' in Christ is profound.

> our work of building for the gospel will not be completed until the Lord Jesus returns

Our grasp of the gospel and of its bearing on life and society always needs to be stretched. We need to learn how to apply its timeless truth to contemporary needs, and to engage the minds and hearts of a new generation. Our work of building for the gospel will not be completed until the Lord Jesus returns.

Paul's letters to the New Testament churches reflect a perceptive grasp of human weakness and of spiritual aspiration. The famous Victorian preacher, Charles Haddon Spurgeon, coined the phrase 'bibline blood'. He wanted his own church family at the Metropolitan Tabernacle in London's Elephant and Castle to get the Bible into their bloodstreams, so to speak; into the way people related to each other. That is the spirit of Paul's writing where biblical values become the 'pulse' of day-to-day living. The 'one anothers' (Appendix 1) express these values as they touch on the way Christians relate to each other. We will always have to keep on working at these things.

39

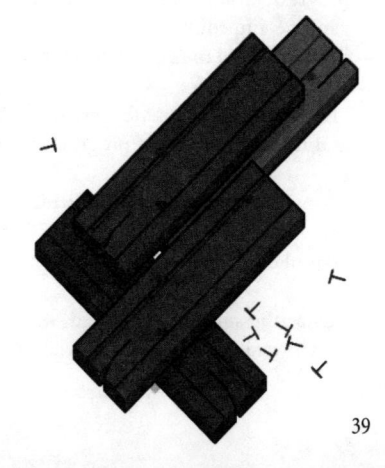

Building projects can be likened to climbing a steep mountain, as Paul Batchelor explained to members of the Round Church in Cambridge (now at St Andrew the Great, better known as StAG). In both, he said, we have a clear objective in view and an unshakeable guide.

· We know the route to take, and we have the best possible back-up team.

· There are obstacles to navigate from the earliest stages, and at times the going can be very tough. We know we can trust our guide, but we still at times experience doubt and fear. Sometimes we feel like giving up and turning back.

· As we look back we snatch a glimpse of the view. We can see some progress has been made. The end seems more worthwhile, and the seemingly 'impossible' becomes the 'very difficult'.

· We must keep the summit in our mind's eye, even when clouds descend. People have made financial sacrifices, and are honouring their promises. We remember this, and we think of the spiritual benefits that reaching the summit - gaining the new facilities - will offer. It spurs us on.

· The summit seems closer, then recedes. We suddenly meet a new difficulty, another long haul, which we had not anticipated.

· We have to accept that not everyone in the group will make equal progress. Doubt can keep creeping back, and for some this will be a genuine impediment. How vital for the leaders to trust the guide implicitly, and to take others with them in that.

· Every so often, pause for rest and refreshment, take a look at the view and remind yourself of progress made. But don't loiter too long.

· For success in climbing a difficult mountain, we need to dig deep into resources of stamina and energy. To keep our eyes on the summit will take all our spiritual reserves.

· Once the summit is reached, we can see other summits to aim for, which were previously not in view.

10
A lot of money!

Faith and finance are intertwined in a story like this. Media pundits may tell us how successful an able orator can be in exhorting people to give. But building projects need serious-minded and long-term giving, and more than oratory is needed. The Holy Spirit must take the message home to people's hearts.

Whoever sows sparingly will also reap sparingly; whoever sows generously will also reap generously. Each man should give what he has decided in his heart to give, not reluctantly or under compulsion, for God loves a cheerful giver. (2 Cor. 9:6,7)

Only the Holy Spirit can convince people of the need for a project to move forward. The whole church membership, or at least the majority of members, have to want it to happen or it will not gain the necessary financial backing.

Giving is a personal matter, serious, and for church members to resolve before God with their families or on their own. People should not feel pressured to give. Many churches adopt the principle that the pastor is not informed how much anyone gives. That knowledge is confidential to the treasurer. In whatever way people respond to the building project - whether enthusiastically, sympathetically, neutrally or antagonistically - the pastor has a responsibility under God to be their pastor, and nothing should be allowed to affect that, or be perceived to affect it.

> faith and finance are intertwined in a story like this

Typically a church will have few if any high earners; more will be in the middle-income group. And then there are pensioners, young singles and newly-married couples, many with more limited incomes; others will be working voluntarily or be unemployed.

We all know the thoughts going through people's minds. How might things be juggled? To see a project through, everyone needs to pull together; to give what they can, indeed to give all they can. The church is a family, and there is no distinction in that regard between those in well-paid professions and those in more modestly paid jobs. Some can give large sums of money; others less. Equally, some are time-rich; others time-poor.

St Paul's & St George's spent five consecutive Sundays looking at New Testament teaching about money, and Jesus' attitude towards money. Only one of those weeks was on giving; the others covered biblical principles of stewardship in general. This was a year before the first pledge day. In this way church members had plenty of time to process and put into practice the biblical principles which they had learned.

As the first pledge day drew nearer, Dave Richards preached a series of sermons on the life of Moses leading the Israelites out of Egypt. This brought good lessons in trusting God for the future, for the unknown.

> An envelope was dropped through the door of one rectory from a woman whose husband was ill, promising £10 per month in cash. To receive notes like that is very humbling. They mean more than the writers know.

Giving early makes a difference

With careful management of funds, compound interest will accrue to a significant sum. Where people come on board early with gifts, there is distinct benefit.

Understandably some hold back. They are not confident that sufficient money will be raised. Once they see the level of giving rise, they too will give. In a major project like this, the balance between those who will give in faith and those who want first to see success round the corner is critical. 'Lord, I believe. Help my unbelief,' must surely be many people's prayer as they make their gifts and fill in their pledges before it becomes definite that permission has been granted for the project to go ahead. And they probably become the major prayer force in asking for others to be moved from scepticism to risk-taking faith.

It is a false hope to think that presently unknown trusts 'out there' will somehow save a congregation from having to give money for building projects. As one church member said, 'This is really just housekeeping. Of course we expect to pay for it ourselves.'

We have the money!

There is a popular cartoon with a minister giving announcements in a Sunday service. It runs like this:

Minister: The good news is that we already have all the money we need for our building work.

(The cartoonist shows the faces of the congregation. Everyone looks pleased, relieved, assured. Someone else has spared them from having to give. That is good news! But things change in the next frame…)

Minister: But the bad news is that it is still in your bank accounts.

We laugh because it is funny. That kind of caricature has a wonderful simplistic charm. But such an approach will never draw the hard cash it is looking for. There is a world of difference between what are sometimes called the 'ethics of guilt' and the 'ethics of gratitude'. If several, even many, people are embarrassed into writing a cheque for two or three hundred pounds, it might mend a leaking roof or replace rotting boards. That is the ethics of guilt in action. Consciences are appeased, and there is a sense of having done one's bit. But the ethics of guilt will never be imaginative and creative. They will not, they cannot, create the fitting new premises for corporate worship and for evangelism which churches up and down the country need today.

Very different are the ethics of gratitude, where giving springs from thankfulness to God, truly believing that all we have comes from him, and being thankful that we ourselves have been bought with a price by Jesus' death on a cross. That makes all the difference in the world. How can we give him enough in return? And how can we not give with joy?

The diagram below shows the breakdown of income from one of the churches featured: the joint effort of those with large incomes, those with middle incomes and those with smaller incomes, of pensioners, and of young people with pocket money. These are spiritual enterprises, not a competition for the size of gift. People must be invited to give cheerfully, and as they are able.

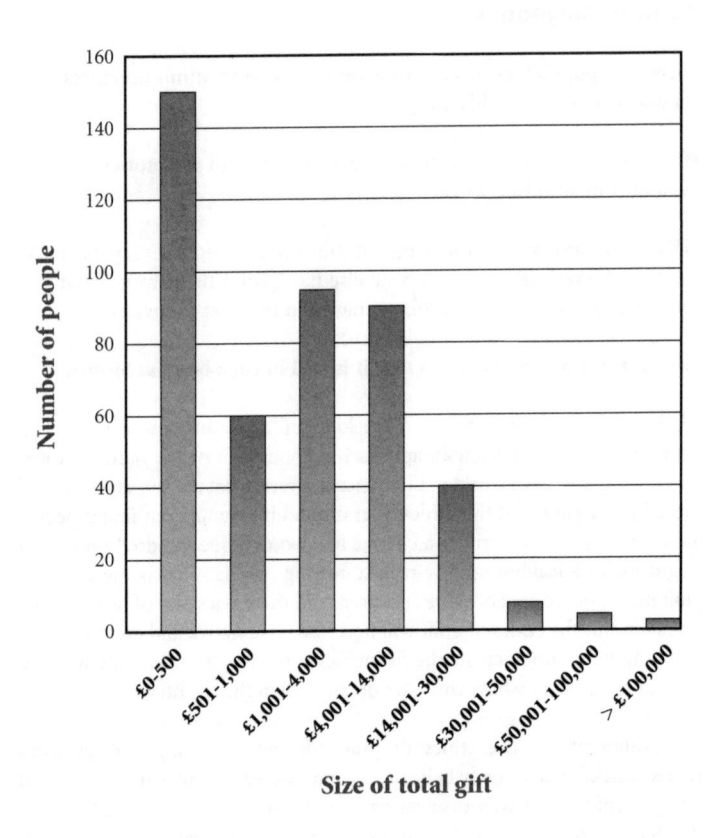

Size of total gift

Jesus is very interested in giving

Jesus is very interested in giving, and in Mark's Gospel we read that he sat himself down right opposite the place in the temple where the offerings were left, and watched the crowd putting their money into the temple treasury.

Many rich people threw in large amounts. But a poor widow came and threw in two very small copper coins, worth only a fraction of a penny. Calling his disciples to him, Jesus said, 'I tell you the truth, this poor widow has put into the treasury more than all the others. They all gave out of their wealth; but she, out of her poverty, put in everything - all she had to live on.' (Mk. 12:41-44)

Some of the most faithful giving in the course of these projects will be from those of limited means who give what they can afford, in cash in envelopes, weekly or monthly, over several years. Motivation is everything in God's sight. These people give out of their love for him.

But creating new buildings will also require a lot of people to give substantial sums 'out of their wealth'. How vitally important it is, in a spiritual sense, that these greatly needed major donors should be giving for the right reasons too.

Passing the million milestone

After an initial appeal, pledges often creep up only slowly, and at times seem to be stuck. Once a certain milestone has been reached (£1m in the case of St Nicholas) the inflow of pledges often stalls. This period can bring feelings of uncertainty and at times, for some, near despair. When that stage was reached at St Nicholas, just over 300 of the 520 members had given. It was a classic point for this to happen; the initial excitement was beginning to wane but it was too early for any visible signs of progress. Detailed designs were still being developed, and would possibly push the initial estimates further up. But the gifts already received and the tax recovered on them did represent a significant financial milestone.

The 'Milestone' brochure arrived in every member's home when some were already beginning to feel the effects of giving-fatigue; others who had been hesitant earlier were even more hesitant now - interest rates had fallen and those with substantial savings had taken an economic beating.

The brochure was written jointly by David Brewster, Legal Director for IMRO (Investment Management Regulatory Organisation), who chaired the Finance Advisory Group, and Paul Batchelor. It laid out the matter clearly:

> God has faithfully met our prayers. He has removed all other obstacles from our path. Now he is challenging us to make some sacrifices, and to return to him part of the gifts he has given to us. We should welcome the challenge.
>
> The time has come to put doubts and division behind us, and to unite in the cause of the One whom we all desire to serve, and whose gospel we wish to proclaim more effectively here in Sevenoaks. It is not a matter for others, but for each and every one of us.
>
> · Of those who have already given generously, we ask, can you do more?
> · Of those who have made a start, we ask, can you go further?
> · Of those who have waited, we ask, is not now the time to start?
> · Of those who have doubted, we ask, can you now see these signs of God's will and join us?
>
> It has perhaps fallen to us to have the privilege to equip St Nicholas for many generations to come. Two years ago we were moved to pledge almost a million pounds for this cause. Most of this is now being faithfully contributed. Please pray that, in a few weeks' time, we will be similarly moved so that, together, with God's help, we can do the same again.

The leaflet closed with the increasingly familiar words of the Apostle Paul to the Christians in Ephesus: 'Now to him who is able to do immeasurably more than all we ask or imagine, according to his power that is at work within us, to him be glory in the church and in Christ Jesus throughout all generations, for ever and ever!' (Eph. 3:20)

Eating elephants

The celebrated Q&A in the Time Manager system runs:

Q: 'How do you eat an elephant?'
A: 'Bit by bit.'

Is that a facile question for executives who conduct their lives around its eight-point diary system? It would seem not. Elephant-sized tasks in any field need to be broken down. It is no different in churches. We must all break down our elephantine tasks and, with God's help, tackle them bit by bit with the human-sized energy he has given us. Hudson Taylor moved from 'Impossible' through 'Difficult' to 'Done' in just this way. He walked up and down Brighton beach on 25 June 1865, praying for '24 skilful, willing workers' to go to China. What faith to ask for two dozen people, who as yet had no Chinese language, to make an impact on a massive country with an ancient culture steeped in 'isms' and folk religion! It sounded impossible. But this was the first step in taking the gospel of Christ to its seemingly impenetrable inland provinces. China was the size of a whole herd of Time Manager's notional elephants, not just one. But two missionaries for each province and two for Mongolia was a start. Similarly, the cost of a building project can seem impossible, but when looked at bit by bit, the figures can come into focus.

Where gospel purposes are in view, many churches will not want to apply for funds from the national lottery. Extra giving is seen as a privilege for church members, a chance for them to express their love of Christ in a tangible way. In Leyland all church members received a personal visit from the churchwarden and his wife, on behalf of the Vision Builders team, specifically to invite a pledge. In Edinburgh and in Sevenoaks the congregations were invited in small groups to hear about the plans over coffee and cake, and to receive a special Giving Pack; this enabled all church members to talk about the project and to think and pray seriously not only about how much to give, but about how and when to give to greatest effect.

As with many churches, the St Nick's congregation included several women whose husbands did not share their faith and commitment. It was not felt right for those wives to put pressure on the family budget for a project like this. Two moving stories of people in this situation were to emerge. One such woman received an unexpected legacy, which she felt able to give without affecting the family budget. Another found herself with a tax repayment she was not anticipating. Again, this was something she could give. These were gifts from the heart. Coming as they did from wives who wanted to observe the biblical pattern of not antagonizing an unconverted husband, they were especially meaningful. It was as if God had provided the means.

Fundraising ideas

Fundraising initiatives can be fun and build community along the way.

One church opened a talent fund with a gift of £200. Anyone could ask for money from the fund to invest in personal moneymaking projects. It could cover, for example, ingredients for baking, or material for sewing. The children could receive money for ingredients to make cakes to sell to the church family.

Here are a few examples of what people have done:

- Cleared attics and sold what they don't need on eBay, or in a local auction sale.
- Offered New Testament Greek lessons, or introductory courses in a modern language.
- Made baby clothes and cushion covers.
- Painted watercolours of churches and sold prints; or accepted commissions to paint watercolours of houses or local scenes.
- Gathered recipes for a church cookbook, always enhanced by contributions from children and teenagers.
- Put on a special concert by the choir.
- Opened their gardens.

Everyone can get involved. Students can be employed as casual gardeners, car washers and decorators when they are at home during the vac, and give a proportion of their earnings to the building fund. Children can make cards and bookmarks to sell. One church gave all the children a tube of Smarties. There was a deal attached: when empty it became a piggy bank for the building project.

While adding to the income streams, fundraising activities also strengthen the church's sense of community. St Paul's & St George's, with a congregation drawing in many students and young professionals, used sports sponsorship in a creative way. One group abseiled off the Forth Road Bridge, sponsored by the metre. Others ran the Edinburgh Marathon, either as individuals or as part of a relay team. They would go out running together in the evenings and draw in other friends outside church to train with them. In a big church like P's and G's some had not met each other before; here was another means of growing and strengthening friendships, within and outside the church family.

Having fun along the way does not detract from the seriousness of giving. Direct giving will always be the major source of income.

'Buy at'

In common with a growing number of charities and churches, P's and G's joined the 'Buy at' scheme. Through this they were able to engage the help of Tesco, Amazon and John Lewis, together with a range of other major UK names. These companies give a percentage of income to a named charity for all online purchases. A link to 'Buy at' was put on the church website. As church members got into the habit of ordering online and naming P's and G's as their chosen charity, an income stream was created which would continue to benefit the church.

Clearing the building

There has been a market for pews to be used in pubs for several years; and church members with larger rooms may like to purchase a short pew, or a length of one, to have in their home. But what else could raise money when a church building is cleared? EBay has no limits. Cast iron radiators could raise a few hundred pounds. We can no longer assume anything is worthless just because we no longer have use for it.

God keeps his promises

Even with a seven-figure sum in the bank, there can be hurdles ahead, especially when a church is dependent on permissions from others outside the situation. This is always the case for Anglican churches, as we have seen, where the diocesan Chancellor holds the authority to grant permission in his gift. He may feel he cannot responsibly rely on the rest of the money coming in. This was the situation in Sevenoaks up to just a few days before the time when the tenders would run out. News of the Chancellor's reluctance to grant permission came on Monday; the tenders would become void by noon on Friday that week.

> a revised business plan was hastily prepared

A revised business plan was hastily prepared. It demonstrated the faithful record of giving over the previous two years; explained how the cash flow would work during construction; and set out the consequences of delays in going ahead. It also chronicled the depth of financial and business experience which members of the Finance Advisory Group brought with them.

But the Chancellor still had questions. With the prospect of costs escalating, or even of having to re-tender, the pressure was on. A further letter was written that Wednesday evening, answering all the questions raised - about the profile of giving, and the number of people who had given to the project. It also sought to allay fears that some might have been coerced into pledging more than they could afford. Janet Batchelor delivered it personally to the Chancellor's home on the Thursday morning.

At 10.00 a.m. on Friday, Miles Thomson received a call. The Diocesan Chancellor had agreed to grant the Faculty. There were only two hours left to spare. God had taken the church family right down to the wire.

That Sunday evening towards the end of the service there was an enormous clap of thunder and simultaneous lightning. As the service closed the arch of a bright rainbow served as a reminder that God keeps his promises.

Effective giving

There are both practical and spiritual lessons to learn from a project like this:

· All gifts are important and valuable. No gift is too small to consider. All promote a sense of involvement and commitment.

· Sustained, sacrificial giving is the most effective. Regular giving becomes part of one's way of life. The sums donated earn interest, and often attract tax relief. The combined effect is very striking. In the St Nick's project, the combined income from interest and tax recoveries was over £500,000.

· The most tax-effective way is giving through the Gift Aid scheme (for lump sums) or through giving agencies like Stewardship or the Charities Aid Foundation (CAF).

· For individuals and families, it is important to plan and budget for giving. It may help to set the sum aside in a separate account. In that way it does not get diverted unintentionally.

· If you cannot afford to give much money, give your abilities, and don't underestimate the value of giving time to prayer.

· Be a cheerful giver.

· Remember that all we have comes from God. We are merely giving back a small part of that.

· God rewards our giving many times over, spiritually.

Also see *The Grace of Giving* by John Stott. Details on p72

11
The Exile

Some churches can move into their hall while the building work takes place. This was the case for St Andrew's, Leyland in its initial phase. Where the hall is part of the building plan, the church has to find accommodation elsewhere, preferably with good public transport links. The congregation from Holyrood Abbey Church moved two miles away to Leith Academy for Sunday services, using nearby Restalrig Church for midweek meetings. P's and G's moved to Pollock Halls of Residence, the university complex on the edge of Holyrood Park. This not only provided a good hall in which to meet, but it also gave a useful opportunity for members of the Christian Union to invite friends to services who might not be so inclined to come to a church.

It is quite common for churches to refer to their time out of the building as being 'in exile'. This is derived from a link (rather tenuous) with the Jews who were taken into exile in Babylon, and eventually allowed to return to Jerusalem to help rebuild the temple. Each church will have stories to tell of these months out of their building, of the new dynamic that this time gave, and of their return. We pick up now on the exile from St Nicholas.

Leaving the church building

Just ten days after the Chancellor's permission had been received for the Undercroft to go ahead, the church family of St Nicholas moved out.

Major refurbishment can entail a long period in exile. The final Sunday in the church building is bound to be a significant day for everyone. Some of the very elderly must wonder if they would worship there again. Enthusiasm for the plans from this sector of the church family in particular means a great deal. They are likely to find it harder than others to worship in a hired hall.

> major refurbishment can mean a long period in exile

As St Nick's moved out, Miles preached from 1 Samuel 7:

'The Philistine army had been routed, and Samuel wanted to mark a great victory. He did so by setting up a large stone, which was like a war memorial, but with one big difference. It didn't contain the names of the dead - those who had died in battle - but just one name, the name of the living, the living God who had helped them to win the victory. As we think of building for the gospel we can say, as Samuel did, "In everything has the Lord helped us".

'That Stone was a powerful reminder. As they looked at it, it would strengthen their trust in the Lord for the next challenge. We can look back and rejoice. We can look ahead and trust because in everything has the Lord helped us too.'

Quoting from Joseph Hart's hymn, Miles finished: 'So we'll praise him for all that is past, and trust him for all that's to come.'

A crèche was arranged for the evening service, so whole families could come if they wanted to. This was an historic day: the culmination of dreams and prayers over 30 years. The final hymn sung in the church before the exile took up from where Miles had finished in the morning. Everyone stood to sing:

How good is the God we adore,
Our faithful, unchangeable friend
Whose love is as great as his power,
And knows neither measure nor end.

The remaining lines were to wait for another 15 minutes. First people all picked up a kneeler to carry it to the back of the church with them. This was a small help for the volunteer force who were to clear the building the next day, but also - and more importantly - a symbolic act. The whole church family was 'hands-on' in that earliest step of transforming the old St Nick's into what they had dreamed of, prayed for, and given for. Then everyone snaked up Six Bells Lane and on down to the church hall, where the singing of the hymn continued:

For Christ is the first and the last;
His Spirit will guide us safe home:
We'll praise him for all that is past
And trust him for all that's to come.

After committing to God in prayer all that lay ahead in this great adventure of faith, a large cake was cut and handed round. It had become a St Nick's tradition for every important occasion to be marked in this way.

Life in exile

From now on, the weekly notice sheet would carry a large number on the top right hand side, with a countdown of the number of weeks to the return. The following Sunday it read: '78'.

Aware that on any given Sunday there might be visitors to a town parish church, Ian Dobbie, the Project Director, arranged a rota of volunteers to stand outside the church and point them in the right direction for the service. One such visitor was a charming Japanese lady. She was visiting London for the first time, and made the journey down to Sevenoaks one Sunday morning, especially to see the church. It wasn't its ancient foundation which had drawn her, nor had she heard about B for the G. But in the 1920s, a missionary called Elsie Baker had been sent out by St Nicholas to work in Osaka. Sometime during her 40 years in East Asia, she had planted the seed of the gospel in this Japanese visitor's life, others had since watered and God had given the growth. This lady's visit was a wonderful reminder of the way the local church is linked with the worldwide church, and of how one woman from a comfortable English parish can have an influence for Christ which will stand the test of time. Shinto-Buddhism, so much a part of Japanese culture, can seem virtually impenetrable.

Almost two years later, the last evening service in the church hall closed with the first verse of the same hymn: 'How good is the God we adore'. The church family had been through two momentous years, and learned more of God's goodness in that time. The reverse process took place, along Six Bells Lane and to the west door of the church. The building was not yet handed back, so it was not possible to go in. Miles hesitated as he was about to announce the start of the second verse. Were 'the Dorothys' there yet?

Dorothy Badman and Dorothy Corke walked a little more slowly than others, and everyone needed to be together for this. The two Dorothys were typical of a large band of elderly people who had been members of St Nick's for decades. Faced with change, they did not react against it, but weighed the arguments. Not only were they accepting of it, but they welcomed it.

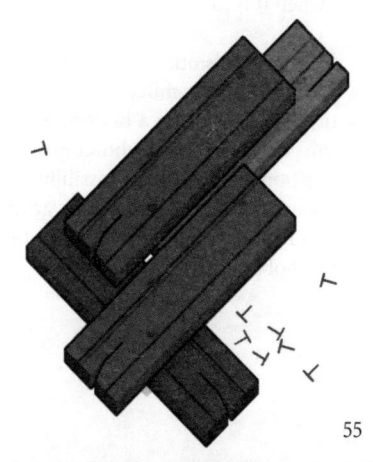

12
Moving back

As we saw in Chapter 4, building projects open up all sorts of possibilities for ancillary use of the premises, and the church family will bring dreams and ideas of ways they would like to see it used: for specialist lectures, concerts, children's activities run by townspeople… These put the church on the map for a wide-ranging public. That can only be good, so long as the administrative and cleaning loads do not become intolerable, and outside use does not crowd out the church's own activities.

In each of the churches featured here, the church council invited a wise and experienced church member to consult leaders of present activities as to their hopes and aspirations, and then to open the same question to everyone with regard to extra uses. They also talked with their counterparts in churches which had already completed such a project. There is much to share, much to learn. To see these wonderful, new God-given premises pressed into their fullest service was everyone's desire.

Getting it right

(1) The building

Interior decor is conspicuous when it is not appropriate, and a joy to the eye when it is just right.

To entrust decisions to a small group who have a good aesthetic sense works well. The architects will provide a steer and the group can work with that, and within a budget, to decide on the final choices. It can be worth visiting several churches and perhaps the Christian Resources Exhibition to research possibilities. For the chairs the St Nicholas ambience team chose a pale terracotta fabric which catches the sun and adds a lightness to the church in a timeless and pleasing way. The carpet is an unobtrusive pale brown.

(2) The people to help

The opening verses of Acts 6 lay out a dilemma and how it was resolved. The church needed extra help: volunteers who could take responsibility for the distribution of food. What kind of people would they look for to fulfil this role? Those who were 'full of the Spirit and wisdom' (v.3).

So much of a church's life is dependent on volunteer staff who handle catering arrangements. As new premises create a range of new uses, the catering becomes a critical part of the church's life. Consistency and a gentle spirit are tested in the church kitchen. For a catering team to be led by someone who has a servant heart, is an encourager by nature, and can see the funny side to everything, is a wonderful tone-setter in any church. Marilynn Sowerby was just like that. For years she led a team of volunteers in St Nick's, always spotting the chance to draw in newcomers, enabling them to get to know others as they prepared food or washed up.[9] Helpers in the church kitchen and helpers in all the midweek activities, as Acts 6 shows, are playing a role which complements that of the pastor and the leadership team. While their time is spent in practical work, it will be done out of love for Christ, and with a desire to serve him.

A growing number of evangelical churches now participate in the interdenominational 9:38 Ministry Training Scheme (taking its name from 'the other Lord's prayer' in Matthew 9:37,38) or its FIEC equivalent, Prepared for Service. These programmes, which last for one or two years, have three aspects. They give new graduates (i) hands-on training in pastoral ministry; (ii) a programme of guided theological study; and (iii) the chance to serve in practical ways to lighten the load on the church staff. Apprentices have much to offer in ministry and can multiply the number of hands available for parish outreach, children's work etc. But just as vitally, this is a time for the church family to invest in them. They will become the next generation of pastors and missionaries.[10]

> **Thanking the work force**
>
> An invitation to lunch for the men who work on the site, together with their families, can be greatly appreciated. Few see the end product of their labours as they move on to other jobs once their skills are no longer necessary. This can be a particularly memorable day for them all. After watching stills or footage of the progress, everyone can sing a hymn in the bright church their labours have created, ideally one which may be familiar from schooldays. Then following a short prayer of thanks for the skills God has given them to complete the task, the pastor can speak briefly, and perhaps give everyone a book to take home as a gift.

Celebration Praise

Here is the climax of the labours exerted over the years - the yearning, the prayer, the giving, the evening committee meetings after a long day's work. The end of that process has come, with the start of all that lies ahead.

The Campaigners formed a Guard of Honour for the Bishop of Blackburn as he entered St Andrew's, Leyland on an October Sunday afternoon to dedicate the refurbished building. The service opened with praise of Christ in the singing of the great Trinitarian hymn dating back to the seventh century:

Christ is made the sure foundation,
Christ the head and corner-stone;
Chosen of the Lord and precious,
Binding all the Church in one;
Holy Zion's help for ever,
And her confidence alone.

It is God who is at work in us to will and to work for his good pleasure. Anything we achieve has been through that work in us. The glory belongs to God and not to us. The service in Leyland closed with the hymn 'To God be the glory'. It was a hymn the church often sang through the project, and those same words are engraved on the bronze plaque commemorating his enabling of the Vision Builders, now mounted at eye level on the south wall of the chancel.

A similar service was held in St Nicholas on a Sunday evening in late June. Richard Bewes, then Rector of All Souls, Langham Place, preached and the All Souls Orchestra came as well, joining forces with the St Nick's musicians.

On that summer's evening, with light streaming through the stained glass windows, the church looked wonderful. Many visitors and local dignitaries had come for the formal opening earlier in the week; now the church family was back home on its own. Richard Bewes turned to the New Testament, and the final chapter of Paul's letter to the Galatians.

An ancient building had been altered for the contemporary world; a medieval church for postmodernity. There was every reason to feel pleased, with all the real and subtle dangers of self-congratulation. The Apostle Paul knew human nature and he cast the right perspective as he looked away, to the cross of Christ.

'May I never boast, except in the cross of our Lord Jesus Christ, through which the world has been crucified to me and I to the world.' (Gal. 6:14)

The words went to the heart of the gospel, with all its paradoxes of death and life, joy and pain. For all that had been achieved in human terms - by those who planned, those who gave money, and those who brought their practical skills - the Undercroft was nothing to boast about. But the cross of Christ was everything to boast about. It was, he said, the epicentre of the Christian faith and the great interpreter of life. By it, and only by it, could values be measured; it was the definer of all choices.

> the cross of Christ is the epicentre of the Christian faith and the great interpreter of life

Handel's 'Hallelujah Chorus' brought everyone to their feet. Here is the compelling truth to drive every church building project. The Lord Jesus, our crucified and risen Saviour, is King of kings and Lord of lords, and he shall reign for ever and ever.

'Now to him who is able'

This setting of Ephesians 3:20,21 was written by Peter Young for the re-opening of St Nicholas. He has kindly made it available for use in worship by any church.

Him be glo - ry in the church ___ and in Christ
Him, Al - le - lu - ia alle lu - ia,
Je - sus Through - out all gene - ra - - - - tions, for
Al — le — lu — ia —,
ev - er and ev - er, A - men
1st. Now to
2nd -men
A - - -
1st
2nd -men.

Afterword

Angus MacLeay became Rector of Sevenoaks in 2001, six years after the Undercroft had been completed. How have the new premises enhanced gospel opportunities?

Having arrived as Rector of St Nicholas after the completion of the Undercroft, it soon became evident to me how significant the project was for the ongoing work of the gospel.

First, the Undercroft provides a wonderful welcoming venue for evangelistic events, especially those which require catering. In any year we find ourselves running Christianity Explored and Discipleship Explored courses as well as other evangelistic events such as jazz suppers, 'Easter in Art' or retired men's lunches. The catering facilities enable us to put on these courses and events in an attractive way.

Secondly, the Undercroft provides opportunity for weekday contact with non-Christians through the usual range of Mums & Toddlers groups and through the bookshop and coffee shop. The historic building at the north end of Sevenoaks is clearly now a seven-day-a-week church with people from the church family and the wider community coming and going at all times each day. Through transforming the building, the evangelistic ministry of the church has also been transformed.

Building for the Gospel equally had an effect on the individual lives of the church family. It has helped to change mindsets and forge a deeper quality of Christian

> sacrificial giving is now part of the lifeblood of the church

discipleship. Although no church family is perfect and we all have our blind spots, the vision of my predecessor, Miles Thomson, left a church family that had learned at first-hand what sacrificial giving entailed. As a result it is now in so many ways characterized by generosity. In the 12 years since the completion of the Undercroft, St Nicholas has given generously to the building project of a partner church in inner-city Liverpool. It has also sent out and funded a steady stream of younger men and women each year into full-time salaried gospel ministry in this country and abroad. Somehow, sacrificial giving is now part of the lifeblood of the church and I am sure that much of this was learnt through the exciting but difficult years of the building project.

So whereas *B for the G* has a continuing evangelistic impact in the parish, its ripples, through the training people receive here and through ongoing sacrificial giving, are reaching to the ends of the earth.

Endnotes

1 Drawing from thinkers and writers including Mother Teresa, Woody Allen, Bertrand Russell and columnists from *The Independent* and *The Economist*, he identifies (i) 'the quest for transcendence', (ii) 'the quest for significance', and (iii) 'the quest for community'. See *The Contemporary Christian* in *The Essential John Stott* (InterVarsity Press, 1999), 521 ff

2 A note for Anglicans. There is some concern that evangelical vicars could possibly be ousted by liberal bishops and so congregations may fear their building work will be only for the short term, should this happen. However, where a parish church has an evangelical patronage, the patrons are able to ensure an evangelical succession. It would be very rare for any church (i.e. the body of Christ's people) to have its preference for evangelical oversight overruled, and for it to be constrained to leave its building.

3 James Hudson Taylor founded the China Inland Mission (now OMF International) on 25 June 1865 with '£10 in the bank and all the promises of God'. There are two major biographies of him, the classic by his son and daughter-in-law, Dr and Mrs Howard Taylor: *James Hudson Taylor: A Biography* (Hodder & Stoughton Religious), and a more recent one *J Hudson Taylor: A Man in Christ* by Roger Steer (Authentic Media).

4 I have drawn from different churches' stories along the way. All of the churches produced similar literature, but I refer only to one example at each stage for easier flow.

5 Here he outlines the basic needs of a church most likely to succeed in a major project. The list begins with being conservative theologically, having strong pastoral leadership and includes a commitment to prayer, well-applied Bible teaching, and generosity in giving. There will be a range of views among evangelicals as to how some areas are best worked out, but the paper, which finishes with advice for any congregations considering a church plant, is stimulating for everyone.

6 Tony Wilmot, already retired, was converted as a schoolboy and joined the St Nicholas youth group, Contact, in the 1930s. He retained his links with the church throughout his many years in Africa as a colonial servant and businessman. He was a man of extraordinary energy. He founded the Nigerian Fellowship of Evangelical Students and was a moving sprit behind the founding of several sister movements (all part of IFES) in African nations. He also founded the Nairobi Evangelical Graduate School of Theology (NEGST). For further information see the *Oxford Dictionary of National Biography* or Lindsay Brown, *Shining like Stars* (InterVarsity Press, 2006) p104.

7 Campaigners is an esteemed national evangelical movement with uniformed groups in many churches.

8 Books on this issue include Mark Greene, *Thank God it's Monday* (Scripture Union Publishing); John Becket, *Mastering Monday* (InterVarsity Press, 2006). Julian Hardyman, Glory Days (InterVarsity Press, 2006) addresses how we use the whole of our lives for God.

9 Marilynn Sowerby also cooked for the speakers and stewards at the Keswick Convention, assisted by a team from other parts of the UK who returned year after year, surely a tribute to her team leadership skills. Marilynn is one of those to whom this book is dedicated. She joined the great cloud of witnesses shortly before Easter 2007.

10 For further information on apprenticeships see www.ninethirtyeight.org and www.fiec.org.uk. Recommended reading for anyone exploring Christian ministry: Vaughan Roberts and Tim Thornborough (eds), *Workers for the harvest field* (Good Book Company, 2006); Robin Wells and Rose Dowsett, *Jesus says Go* (IFES/Monarch Books, 2006); Ajith Fernando, *An authentic servant* (IFES/OMF, 2006). All are available from www.10ofthose.com

APPENDICES

APPENDIX 1
The art of belonging
The 'one anothers' in Scripture

The emphasis on church as people is brought out by the Lord Jesus, and by the Apostle Paul. As fellow Christians, we belong to each other; our faith is a corporate faith. We are told to encourage one another, counsel one another, love one another, bear one another's burdens. It may seem trivial to move from exhortations like that to the benefits of talking over coffee after the Sunday services, but in a truly practical and down-to-earth way, good meeting facilities help a congregation to be built into a 'church family'. Newcomers can be welcomed there, new friendships can start, and existing friendships grow. As Christians we are part of the Body of Christ. We need each other.

Mark 9:50	Be at peace with each other
John 13:34	A new command I give to you: Love one another
Romans 1:11,12	I long to see you… that you and I may be mutually encouraged by each other's faith
Romans 12:10	Be devoted to one another in brotherly love. Honour one another above yourselves
Romans 12:16	Live in harmony with one another
Romans 13:8	Let no debt remain outstanding, except the continuing debt to love one another
Romans 15:7	Accept one another, just as Christ accepted you
Galatians 6:2	Carry each other's burdens, and in this way you will fulfil the law of Christ
Ephesians 4:32	Be kind and compassionate to one another, forgiving each other, just as in Christ God forgave you
Philippians 2:1-4	If you have any encouragement in being united with Christ… each of you should look not only to your own interests, but to the interests of others
Colossians 3:16	Let the Word of Christ dwell in you richly as you teach and admonish one another with all wisdom
Hebrews 3:13	Encourage one another daily, as long as it is called Today
Hebrews 10:24,25	Let us consider how we may spur one another on towards love and good deeds. Let us not give up meeting together, as some are in the habit of doing, but let us encourage one another
1 Peter 4:8,9	Above all, love each other deeply, because love covers over a multitude of sins. Offer hospitality to one another… Each one should use whatever gift he has received to serve others

APPENDIX 2
Lessons from Nehemiah

Nehemiah has been described as the book for those who want to achieve great things. The St Nicholas staff team preached through it twice over the course of B for the G, once on Sunday mornings, and once on Sunday evenings. It brought help for those at the sharp end, and for the whole church family. The emphasis throughout is of working together. In one chapter alone, as Nehemiah describes the position of people rebuilding the wall, the words 'next to him' appear over 20 times. The detail of the record shows that each person's role was vital. No one's contribution went unnoticed.

Nehemiah led God's people through a remarkable building programme in the sixth century BC, rebuilding a people as well as a city. This book - hidden away in the middle of the Old Testament - provides some key principles for any church in building for the gospel. These principles come in pairs, balancing one another.

Prayer and commitment. While the project grew out of his prayers, Nehemiah was willing to be the answer to his prayers (Chapter 1). Hudson Taylor's son said of his father, 'He prayed about things as if everything depended upon the praying. But he worked also, as if everything depended on the working.'

Vision and planning. Nehemiah was spurred on by a vision; at the same time he did his homework, researching, surveying the situation, and making plans (Chapter 2).

Leader and members. Nehemiah was a team player and he involved all God's people in the project (Chapter 3).

Sword and spade. There were enemies around who were out to stop the rebuilding. Some opposition came from outside and some from inside, as God's people began to grumble under the pressure of the project. But nothing would be allowed to stop the work going forward. So they took a sword in one hand and a spade in the other (Chapters 4-6).

Walls and people. Rebuilding the walls was the prelude to rebuilding the people in a new commitment to their Lord (Chapters 7-13).

Appendix 3
Churches which would welcome visits

The following are just a few churches to have completed building projects in recent years. If you are visiting in the area, you would be welcome to join them for worship.*

SCOTLAND
St Paul's & St George's, **Edinburgh**	www.pandgchurch.org.uk
Holyrood Abbey Church, **Edinburgh**	www.holyroodabbey.org
Sandyford Henderson Memorial Church, **Glasgow**	www.sandyfordhenderson.co.uk
Chryston Church, **Glasgow**	email: chrystonchurch@hotmail.com

WALES
Highfields Church, **Cardiff**	www.highfieldschurch.co.uk
Glenwood Community Church, **Cardiff**	www.glenwoodchurch.org
Beacon Church, **St Mellons**	www.beaconcentre.co.uk

NORTHERN IRELAND
Buckna Presbyterian Church, **Broughshane**	www.buckna.org
Magherafelt Baptist Church, **Magherafelt**	www.magherafeltbaptist.org
Portstewart Baptist Church, **Portstewart**	www.portstewartbaptist.co.uk

REPUBLIC OF IRELAND
Grosvenor Baptist Church, **Dublin**	www.grosvenorbaptist.org
CORE St Catherine's, **Dublin**	www.corechurch.ie

ENGLAND
Church of the Saviour, **Blackburn**	www.the-redeemer.org.uk
St Bartholomew's, **Blackburn**	www.the-redeemer.org.uk
Kensington Baptist, **Bristol**	www.kenbaptist.org
St Andrew the Great, **Cambridge**	www.stag.org
St Paul's, Howell Hill, **Cheam**	www.saintpauls.co.uk
Central Baptist, **Chelmsford**	www.centralbaptistchelmsford.org.uk
The King's Centre, **Chessington**	www.thekingscentre.org.uk
St Luke's, **Cranham**	www.stlukescranham.co.uk
Cranleigh Baptist, **Cranleigh**	www.cranleigh.org.uk
All Saints, **Crowborough**	www.allsaintscrowborough.org
West Street, **Dunstable**	www.weststreet.org.uk
Enfield Baptist Church, **Enfield**	www.babuike.f2s.com

---

* Times of services are on their websites. The building projects range from several thousands of pounds to millions of pounds; and from two- or three-stage projects to major single projects. Some have been carried out in partnership with a local school, or with the borough council. All have been ventures of faith.

St Matthew's, **Fulham**	www.stmf.org.uk
St James, **Gerrards Cross**	www.saintjames.org.uk
St John's, Downshire Hill, **Hampstead**	www.sjdh.org
Burlington Baptist, **Ipswich**	www.burlingtonbaptist.org.uk
St Andrew's, **Leyland**	www.standrewsleyland.org.uk
Headington Baptist, **Oxford**	www.hbc-oxford.org.uk
Oxford Community Church, **Oxford**	www.occ.org.uk
St Aldate's, **Oxford**	www.staldates.org.uk
St Ebbe's, **Oxford**	www.stebbes.org.uk
St Nicholas, **Sevenoaks**	www.stnicholas-sevenoaks.org
New Community Church, **Sidcup**	www.newcommunitychurch.org.uk
Grace Baptist Church, **Southport**	www.gracebaptist.org.uk
St John's, **Tunbridge Wells**	www.saint-johns.info
St Luke's, **Watford**	www.stlukeswatford.org

> If we may add your church to this list for the next printing, please write to the publisher.

PROCLAIMING CHRIST IN THE WORLD'S UNIVERSITIES

In human terms, universities wield the most profound influence in every nation. As Charles Malik, former President of the UN General Assembly once said: 'Change the university and you change the world.'

The International Fellowship of Evangelical Students (IFES) is working to see a clear gospel witness firmly established in every university in the world. We long for more and more students in these centres of learning to meet with the living Christ 'in whom are hidden all the treasures of wisdom and knowledge'. We now have a presence in over 150 countries but some of the toughest frontiers still lie ahead.

We are equipping students to be (i) effective evangelists; (ii) serious disciples; (iii) mission-minded Christians. We want our graduates to help strengthen the worldwide Church, and to bring the living Christ into their professions, family life and society.

To quote Charles Malik again: 'The Church can render no greater service to itself or to the cause of the gospel than to try to recapture the universities for Christ.' Opportunities to bring this gospel to students are greater than ever. If you or your church would like to know more, or to receive prayer news, email **info@ifesworld.org**. We'd love to hear from you.

MORE IFES was formed in 1947 by evangelical student movements in ten countries. (UCCF is a founder member.) Leaders covenanted on behalf of their students and staff to work and pray to see a witness to Christ among students in all nations. That commitment is still strong, right across the world. UK graduates have pioneered national movements in Africa and East Asia, Europe and Eurasia; Chadian students, evacuated during the civil war to study in other African countries, founded IFES movements in eight new nations; Japanese graduates are establishing a movement in Cambodia; Canadians and Filipinos in Central Asia; Koreans and Americans in Mongolia; Ukrainians in the former Soviet republics; Middle Eastern graduates in the Gulf states...

Our movement in the UK is UCCF (formerly the Inter-Varsity Fellowship, and a founder member). Many movements are known as FES, but we are also known, for example, as InterVarsity in North America; GBU in Latin America, Francophone Africa and Europe; CCX in Ukraine, Russia, Belarus; TSCF in NZ.

For more details, go to **www.ifesworld.org** or order *Shining like Stars* by Lindsay Brown from **www.ifesworld.org/books**

10 OF THOSE.COM
THE best way to buy CHRISTIAN books in bulk

For more copies of this booklet go to **www.10ofthose.com**

We carry a wide selection of books – theology, biography, Christian lifestyle, Bibles and commentaries all at guaranteed low prices.

We service many church bookstalls around the UK and can help your church to set up a bookstall.

Single and bulk purchases welcome. For more information contact:
quote@10ofthose.com

10ofthose.com is a not-for-profit limited company.

GIVING: On the subject of personal giving, we recommend *The Grace of Giving* by John Stott (*The Didasko Files* booklet series) ISBN 978-1-899464-01-2 (IFES/Langham Partnership). Order through our site at bulk discount for churches.

www.10ofthose.com